T0298360

Zero to Hero

Zero to Hero: Your Guide to a Career in Cybersecurity is an essential roadmap for anyone aiming to penetrate the vibrant and ever-expanding domain of cybersecurity. In an era where digital threats loom larger and more complex than ever, this book stands as a beacon of clarity and practical wisdom. Tailored for novices and those with basic understanding, this resource empowers learners to solidify their cybersecurity foundation. It stands out with its laser focus on real-world applicability, ensuring readers grasp theoretical concepts and can implement them effectively.

Key Features of This Guide:

- **Actionable Learning:** Dive into engaging exercises, compelling case studies, and practical scenarios that demystify complex cybersecurity concepts.
- **Career Development:** Gain invaluable insights into crafting a standout resume, navigating job interviews with confidence, and learning strategies for a successful job hunt in the cybersecurity realm.
- **Cutting-Edge Knowledge:** Stay ahead of the curve with detailed explorations of the latest cybersecurity trends, tools, and technologies that are shaping the future of digital security.
- **In-Depth Discussions:** From ethical hacking to digital forensics, explore the breadth and depth of the cybersecurity field, ensuring a comprehensive understanding of various career paths.
- **Progressive Skill-Building:** Embark on a structured learning journey, from foundational concepts to advanced techniques, tailored to foster a deep, actionable understanding of cybersecurity.

Zero to Hero: Your Guide to a Career in Cybersecurity is your launchpad into the heart of the industry. Perfect for students, career changers, and IT professionals, this book provides the essential knowledge and skills to secure a rewarding career in this critical field.

Begin your journey from novice to expert in cybersecurity today!

ZERO TO HERO

Your Guide to a Career in Cybersecurity

Felix Kyei Asare
C|CISO, CISSP, CEH, CISM

CRC Press is an imprint of the
Taylor & Francis Group, an **informa** business

Designed cover image: © Emmanuel Asare & Prince Attah

First edition published 2025
by CRC Press
2385 NW Executive Center Drive, Suite 320, Boca Raton FL 33431

and by CRC Press
4 Park Square, Milton Park, Abingdon, Oxon, OX14 4RN

CRC Press is an imprint of Taylor & Francis Group, LLC

Library of Congress Cataloging-in-Publication Data
Names: Asare, Felix Kyei, author.
Title: Zero to hero : your guide to a career in cybersecurity / Felix Kyei Asare.
Description: First edition. | Boca Raton : CRC Press, 2025. | Includes bibliographical references and index.
Identifiers: LCCN 2024022523 (print) | LCCN 2024022524 (ebook) | ISBN 9781032802213 (hbk) | ISBN 9781032818351 (pbk) | ISBN 9781003501589 (ebk)
Subjects: LCSH: Computer security--Vocational guidance.
Classification: LCC QA76.9.A25 A8243 2025 (print) | LCC QA76.9.A25 (ebook) | DDC 005.8023--dc23/eng/20240813
LC record available at https://lccn.loc.gov/2024022523
LC ebook record available at https://lccn.loc.gov/2024022524

ISBN: 978-1-032-80221-3 (hbk)
ISBN: 978-1-032-81835-1 (pbk)
ISBN: 978-1-003-50158-9 (ebk)

DOI: 10.1201/9781003501589

Typeset in Caslon
by SPi Technologies India Pvt Ltd (Straive)

To my Ariya and Kairo
- All I do is for you both.

Contents

About the Author

At the forefront of cybersecurity, **Felix Kyei Asare** embodies the fusion of seasoned leadership with pioneering innovation. With a distinguished 15-year journey navigating the complex terrains of security and technology, Felix's career is a testament to his unwavering dedication and strategic brilliance. His leadership has been instrumental in steering high-performing global teams, managing up to 40 professionals, and spearheading cybersecurity initiatives across a diverse array of industries including financial services, healthcare, and media.

Felix's expertise lies in his versatile approach to cybersecurity, adeptly balancing the demands of rapid-growth environments with the rigors of global enterprise operations. His approach to security transcends traditional boundaries, viewing it as a cornerstone of strategic business growth and sustainability. Through his roles at eminent organizations such as Putnam Investments and Allianz North America, Felix has demonstrated an unparalleled ability to translate complex IT and security challenges into robust, accessible solutions.

Holding an MSc in Cybersecurity and a BSc in Computer Information Systems, supplemented by prestigious certifications from Yale, Carnegie Mellon University, and others, Felix's academic and professional credentials solidify his authority in the field. His thought leadership extends beyond the corporate sphere, enriching the cybersecurity community as a sought-after speaker and mentor.

Felix's commitment to shaping the future of cybersecurity is evident in his proactive efforts to mentor the next generation, particularly focusing on diversifying the cybersecurity workforce. His boot camp program reflects this dedication, equipping aspiring professionals with vital skills and knowledge.

Author's Note:
In the pages that follow, Felix invites readers into the evolving world of cybersecurity through his experienced lens. His narrative is not merely a collection of best practices but a compelling vision for integrating cybersecurity into the fabric of modern business strategy. As a guide, mentor, and leader, Felix's insights offer invaluable perspectives for anyone looking to navigate the complexities of cybersecurity in today's digital landscape.

Specializations include:

- Strategic Development and Leadership in Information Security
- Advanced Information Security Operations and Global Strategy
- Expertise in Incident Response, Threat Management, and Vulnerability Assessment
- Innovation in Fraud Mitigation, Risk, and Third-Party Risk Management
- Comprehensive Knowledge in Information Security Policy, Application Security, and System Implementation

Felix is positioned not only as an expert in his field but also as a visionary leader prepared to address the cybersecurity challenges of today and tomorrow.

Acknowledgments

Writing *Zero to Hero: Your Guide to a Career in Cybersecurity* has been an incredible journey, made possible by the support of many individuals.

I extend my deepest gratitude to my parents, Dr. and Mrs. Asare Yeboah, for instilling in me the values of resilience and faith. To my partner, Whitney, thank you for challenging my thoughts at the airport when the idea for this book came to me, and for managing our home during the countless hours I spent developing it. To my siblings, Alberta, Eunice, and Emmanuel, your unwavering support has been invaluable, keeping me motivated through late nights and challenges.

I am deeply grateful to my college professors Rick Ross and Rob Walsh for reminding me that anything is possible. To my mentor, Kweku Bankah, and coach Ann Holm, thank you for helping me build resilience. To my colleagues from the Charlotte CISO group, your guidance and willingness to share expertise have significantly shaped this book.

To my desig team, Prince Owusu-Attah and Emmanuel (Zone Studios), thank you for your creative insights and for the experiences we've shared in the creative industry.

I extend my heartfelt thanks to Barbara Fant, Henry Mensah, Emmanuel Owusu, Derek Turner, Andres Andreu, Nikita Jones, J Aryeetey, Charity Nix, J Grant, Aashesh Verma, Gaurav Bhardwaj,

J Yeboah, Nsafoah Owusu, and Kathy Saunders for your crucial contributions during the writing and editing process.

I also appreciate the team at CRC Press, especially Gabriella Williams, for believing in this project and bringing it to life.

I am indebted to the broader cybersecurity community, whose ongoing work continues to advance our field.

Lastly, to all the aspiring cybersecurity professionals who inspired this book—your passion for this vital field motivates me daily.

Thank you all for helping transform this idea into a resource that I hope will guide many from zero to hero in cybersecurity.

Felix Kyei Asare (C|CISO, CISSP, CISM)

Glossary

Antivirus Software: A program that detects and removes malware.

Authentication: Verifying the identity of a user through passwords, biometrics, and so on.

Authorization: Defining what actions a user is permitted to perform in a system.

Availability: Ensuring information and systems are accessible when needed.

Black Hat Hacker: A hacker who violates computer security for personal gain or maliciousness.

Compliance: Adhering to laws, regulations, policies, and standards that govern cybersecurity.

Confidentiality: Protecting private or sensitive information from unauthorized access.

Cyberattack: An attempt to access, damage, or disrupt a computer network or system.

Cybersecurity: The practice of protecting computers, networks, programs, and data from unauthorized access or attacks.

Cyber Threat: Any potential danger to information systems and data from cyberattacks.

Data Breach: A security incident in which information is accessed without authorization.

Denial-of-Service Attack (DoS): An attack meant to shut down a machine or network, making it inaccessible to its intended users.

Distributed Denial-of-Service Attack (DDoS): A type of DoS attack where multiple compromised systems are used to target a single system, causing a Denial of Service.

End-to-End Encryption (E2EE): A method of secure communication that prevents third parties from accessing data while it's transferred from one end system to another.

Encryption: Encoding data to prevent unauthorized access during storage or transmission.

Firewall: A network security tool that monitors traffic and blocks threats.

Grey Hat Hacker: A hacker who is in between white and black hats. They may violate ethical standards or principles, but without the malicious intent ascribed to black hat hackers.

Identification: The process of a user providing a unique ID like a username.

Incident Response: The procedures followed to address a cyberattack or breach.

Information Technology (IT): The use of computers, software, and technology systems to store, process, and communicate information.

Integrity: Safeguarding the accuracy and completeness of data and systems.

Intrusion Detection System (IDS): A device or software application that monitors a network or systems for malicious activity or policy violations.

Intrusion Prevention System (IPS): An extension of IDS that not only detects potentially malicious activity but also prevents it.

Malware: Malicious software designed to harm, exploit, or otherwise perform unauthorized actions on a computer system.

Patch Management: The process of distributing and applying updates to software. These patches are often necessary to correct errors (known as "vulnerabilities" or "bugs") in the software.

Phishing: A method of trying to gather personal information using deceptive e-mails and websites.

Penetration Testing: Legally simulating cyberattacks to find vulnerabilities.

Ransomware: A type of malicious software designed to block access to a computer system until a sum of money is paid.

Risk: The likelihood and impact of a cyber threat occurring.

Security Audit: A systematic evaluation of the security of a company's information system by measuring how well it conforms to a set of established criteria.

Security Information and Event Management (SIEM): Provides real-time analysis of security alerts generated by applications and network hardware.

Security Operations Center (SOC): A facility to monitor, analyze, and respond to cybersecurity threats.

Social Engineering: The art of manipulating people so they give up confidential information.

Threat Intelligence: Gathering data to understand cyber threats and adversaries.

Two-Factor Authentication (2FA): A security process in which the user provides two different authentication factors to verify themselves.

Virtual Private Network (VPN): A service that allows you to connect to the internet via a server run by a VPN provider, enabling a secure connection.

Vulnerability: A weakness in a network or system that can be exploited.

White Hat Hacker: An ethical computer hacker who specializes in penetration testing and other testing methodologies to ensure the security of an organization's information systems.

Zero-Day Exploit: A cyberattack that occurs on the same day a weakness is discovered in software. At this point, the software developers have had zero days to fix the flaw.

INTRODUCTION

Welcome, and congratulations on choosing to embark on this exhilarating journey toward becoming a cybersecurity professional! Your decision to open this book tells me you've got the curiosity that is key in our field. This curiosity is often the making of champions in the ever-evolving world of cybersecurity.

In this era of digital dominance, where every organization, regardless of its size, leans heavily on technology, there's an escalating demand for cybersecurity professionals who are skilled, inquisitive, and dedicated. If you've ever felt the field of cybersecurity to be complicated, abstract, or daunting, I want you to take a moment, breathe deeply, and get ready to discard these beliefs. You're on the brink of an exciting, fulfilling journey that will demonstrate just how approachable this field can be.

In the pages of this book, I've taken the task of simplifying the complexities and misunderstandings related to cybersecurity. I've avoided fluff and jargon, instead focusing on clear, simple language accompanied by practical, relatable examples. You won't feel lost in the maze of tech-talk; instead, you'll discover intriguing insights into how businesses operate, the central role of IT, and the increasing threat of cyberattacks.

Together, we'll journey into the expansive world of cybersecurity, unveiling its key facets and concepts, understanding the importance of the CIA triad (Confidentiality, Integrity, and Availability), delving into diverse security measures, and getting familiar with the tools commonly used in the industry. We'll also decode the importance of Regulatory Frameworks and dive into Risk Management, converting abstract ideas into comprehensible, tangible knowledge.

This book isn't just about understanding cybersecurity; it's about equipping you with the right skills to get you from Zero to Hero. After reading this book, you should be able to look at any cyber job posting, pick the language in the job posting apart, and wow recruiters and hiring managers with your skills and composure. You are

DOI: 10.1201/9781003501589-1

going to learn how to craft your resume, how to search for jobs, how to prepare for interviews, what to say and not say in the interview, and even how to negotiate the several job offers that will come your way.

Remember, this journey is yours, and each step you take, including the act of reading this very sentence, brings you closer to your goal of becoming a cybersecurity professional. Prepare to enjoy this journey, challenge your preconceptions, and before you know it, you'll have moved from beginner to advanced level.

Are you ready?

Not so fast!! Before we begin exploring the world of cybersecurity, let's take a moment to gauge your existing knowledge.

Below is a ten-question pre-assessment quiz that covers some of the core concepts we'll be discussing in this book. These questions will help you identify areas you're already familiar with and topics you want to focus on as you progress through the material.

The pre-assessment is meant to be a measurement of where you are starting from – it's not expected you'll know all the answers yet! Be sure to read each question carefully, then select the best answer based on your current knowledge. Go with your gut instinct if unsure. Once you complete the pre-assessment, you'll have a baseline to compare against as you learn these new skills.

According to Cyberseek.org – From May 2022 through April 2023, there were 159,000 openings for Information Security Analysts and 163,000 workers currently employed in those positions – an annual talent surplus of 4,000 workers for cybersecurity's largest job.

Cybersecurity workers

There are 490,513 additional openings requesting cybersecurity-related skills, and employers are struggling to find workers who possess them. On average, cybersecurity roles take 21% longer to fill than other IT jobs.

Pre-Assessment Quiz

Before we jump into the nitty-gritty of cybersecurity, let's take a quick pulse check with this pre-assessment quiz. Think of it as your starting point—a way to see where your knowledge is right now. Don't worry if you're feeling a bit rusty or if some of the questions seem like they're written in binary. This quiz isn't about passing or failing; it's about giving you a baseline. So, grab your virtual notepad, channel your inner cybersecurity hero, and let's see what you already know!

1. What does IT stand for?
   (a) Information Technology
   (b) Internet Transmission
   (c) Interactive Training
   (d) Internal Terminal
2. What does CIA stand for in cybersecurity?
   (a) Central Intelligence Agency
   (b) Confidentiality, Integrity, Availability
   (c) Certified Information Analyst
   (d) Cybersecurity Intelligence Agency
3. What is a cyber threat?
   (a) A computer virus
   (b) Any potential danger to information systems
   (c) Hackers trying to breach security
   (d) Messages from anonymous senders
4. What is reconnaissance in cybersecurity?
   (a) Studying and analyzing a target system
   (b) Disabling computer firewalls
   (c) Publishing confidential data publicly
   (d) Identifying network IP addresses
5. What does QWERTY stand for?
   (a) A keyboard layout
   (b) A password encryption formula
   (c) A cyberattack technique
   (d) An operating system term
6. What is encryption?
   (a) Encoding data for secure transmission
   (b) Decoding intercepted network packets
   (c) Bypassing firewall security controls
   (d) Allowing unauthorized file downloads
7. What is a DDoS attack?
   (a) Distributed Data on Server attack
   (b) Domain Database of Service attack
   (c) Distributed Denial of Service attack
   (d) Dedicated Denial of Sender attack

8. What is ransomware?
    (a) Malware that holds systems hostage until ransom paid
    (b) Software that replicates itself across networks
    (c) Viruses that breach firewalls and APIs
    (d) Worms that exploit operating system flaws
9. What is a proxy server?
    (a) A network gateway that stores cached webpages
    (b) A protocol that allows remote desktop access
    (c) An attack vector used to intercept traffic
    (d) A virtual server located on the deep web
10. What is a vulnerability assessment?
    (a) Scanning systems and networks for weaknesses
    (b) Developing code to exploit weaknesses
    (c) Publishing confidential system data publicly
    (d) Selling access to compromised systems

Answers to the Pre-Assessment Quiz

1 What does IT stand for?
  Answer: a) Information Technology
2 What does CIA stand for in cybersecurity?
  Answer: b) Confidentiality, Integrity, Availability
3 What is a cyber threat?
  Answer: b) Any potential danger to information systems
4 What is reconnaissance in cybersecurity?
  Answer: a) Studying and analyzing a target system
5 What does QWERTY stand for?
  Answer: a) A keyboard layout
6 What is encryption?
  Answer: a) Encoding data for secure transmission
7 What is a DDoS attack?
  Answer: c) Distributed Denial of Service attack
8 What is ransomware?
  Answer: a) Malware that holds systems hostage until ransom paid
9 What is a proxy server?
  Answer: a) A network gateway that stores cached webpages
10 What is a vulnerability assessment?
  Answer: a) Scanning systems and networks for weaknesses

Summary of the Pre-Assessment Quiz

How did it go? Whether you breezed through or hit a few bumps, this quiz gave you a snapshot of your current cybersecurity know-how. If some of the questions had you scratching your head, don't sweat it—that's what this book is here for! As you work through the chapters, you'll fill in those gaps, and by the end, you'll be acing these questions like a pro. Remember, this journey from zero to hero is all about growth, so take note of what you need to focus on, and let's get ready to dive deeper into the world of cybersecurity!

Chapter Breakdown: Mapping Out Your Cybersecurity Adventure

Now that you've got a sense of where you're starting from, it's time to chart the course for your journey ahead. Each chapter in this book is a stepping stone, carefully designed to build your skills and knowledge, piece by piece. Whether you're a newbie or looking to sharpen your existing skills, this chapter breakdown will serve as your trusty guide, showing you what to expect and how each chapter will help you get closer to becoming a cybersecurity hero. Ready to map out your adventure? Let's dive into the details and see what's in store!

Chapter 1: Understanding Businesses: What and Why
Learn about what constitutes a business and why businesses exist. Understand how they generate revenue, their basic functions, and their role in the economy.

Chapter 2: The Role of IT in Business
Explore the importance of Information Technology (IT) in modern businesses. Learn about how IT has revolutionized businesses, making them more efficient, connected, and globally competitive.

Chapter 3: The Pillars of IT and Cybersecurity
Understand the evolution of IT and the main pillars that paved way for Cybersecurity to become a household name.

Chapter 4: Demystifying Cybersecurity: Key Areas and Concepts
Dive into the various areas of cybersecurity. Discover the importance of the CIA triad (Confidentiality, Integrity, Availability), explore different security measures, and familiarize yourself with common tools used in the field.

Chapter 5: Preparing for the Job Hunt: What You Need to Know
Unpack the complexities of job interviews in the cybersecurity field. Prepare for scenario-based and technical questions and learn how to ask insightful questions of your own.

Chapter 6: Mastering Your Resume
Get in-depth advice on tailoring your resume for cybersecurity roles. Learn about using active and passive voice, including metrics, and how to structure your resume to showcase your skills and experience most effectively.

Chapter 7: Understanding the Recruitment Process
Learn about the stages in the recruitment process for cybersecurity roles. Discover what recruiters look for in candidates and how to position yourself as an ideal candidate.

Chapter 8: When You Get the Offer
This final chapter will guide you through the process of receiving a job offer. Learn when and how to accept the offer, understand when to consider a counter-offer, and prepare for the exciting step of beginning your career in cybersecurity.

By the end of this book, you will have gained an in-depth understanding of the cybersecurity landscape, the role of IT in businesses, and the skills and knowledge needed to land your first job in the field. With practical tips and real-world advice, this guide is your stepping stone toward a rewarding career in cybersecurity.

The fastest-growing type of cybercrime is expected to attack a business, a consumer, or a device every two seconds by 2031.

– Steve Morgan, E (Cybercrime magazine)

1

Understanding Businesses

What and Why

1.1 Introduction

Welcome to the exciting world of cybersecurity! No matter where you're starting from—whether you're a complete beginner or have some experience under your belt—I'm here to guide you on a journey toward one of the most in-demand and rewarding careers out there. To start, it's essential to understand the core subject of our protection efforts – **THE BUSINESS**. But what is a business, and why do businesses exist? Where does cybersecurity come into play? Why is everyone talking about this new field?

1.2 What is a Business?

A business is an organization that provides goods or services to customers. These goods could be anything from clothes to electronics, while services could be anything from haircuts to home repairs. Businesses exist in all shapes and sizes, from small local shops to giant multinational corporations. Some of the common businesses you are familiar with are Amazon, Walmart, Delta Airlines, Hulu, Netflix, Google, YouTube, Apple. Shall I continue? I'm sure you get the gist.

1.3 Why Do Businesses Exist?

The primary reason businesses exist is **TO MAKE A PROFIT**. If you don't remember anything I tell you in this book, remember a business's number one priority is **TO MAKE A PROFIT**. They do this by selling their goods or services at a price higher than the cost to produce or provide them.

DOI: 10.1201/9781003501589-2

- **Retail Industry:** Businesses in the retail industry sell consumer goods. A company like Walmart, for example, earns its revenue by purchasing goods from manufacturers or wholesalers at a lower price and selling them to consumers at a higher price. The difference, known as the "margin", is how they make a profit.
- **Technology Industry:** In the technology industry, companies like Microsoft and Apple make money by selling hardware (like computers and phones), software (like operating systems and apps), and services (like cloud storage and technical support). Some tech companies, like Google and Facebook, earn a significant portion of their revenue through advertising.
- **Healthcare Industry:** Hospitals and clinics generate revenue by providing medical services. They might charge patients directly, or they might bill insurance companies. Pharmaceutical companies make money by developing and selling drugs.
- **Financial Services Industry:** Banks and credit unions make money by lending out deposits at higher interest rates than they give to their depositors. They also generate revenue from fees for various services. Investment firms earn money from management fees and commissions on transactions.
- **Food and Beverage Industry:** Restaurants make money by selling prepared food and drinks. They buy raw ingredients at a wholesale price, prepare meals, and sell them at a price that covers the cost of ingredients, labor, overhead, and hopefully a profit.

For example, it costs Apple right around $450 to make an iPhone 14; yet they sell this product for over $1,100. In that same way, a seller on Amazon can buy a fidget spinner from China for 50 cents and sell it on Amazon for $15.00.

Let's do a simple breakdown to understand some key terminology.

In this scenario, we're going to find cost of goods sold, revenue, and profit.

If Apple sells let's just say 2,000 phones in one month, what will their total revenue for that month be?

Number of devices: 2,000
Per unit cost of device: $450
Selling price: $1,100

For 2,000 phones, it will cost: 2,000 × $450 = $900,000

Calculate revenue

$$1{,}100 \times 2{,}000 = \$2{,}200{,}000.$$

The difference between what it costs to produce a good or a service and what customers pay for it is known as **PROFIT**.

Calculate profit

$$\$2{,}200{,}000 - 900{,}000 = \mathbf{1{,}300{,}000}$$

Ransomware will cost its victims around $265 billion (USD) annually by 2031, Cybersecurity Ventures predicts.

1.4 Why Understanding Businesses is Important

Regardless of the specific ways these industries generate revenue, the ability to do so effectively determines the health and viability of a business. If revenue declines, businesses may have to cut costs, which could involve reducing their workforce, halting expansion plans, or even closing down in severe cases.

In the realm of cybersecurity, understanding business fundamentals and revenue streams is paramount. A well-executed cyberattack can cripple an organization's ability to generate income, leading to catastrophic financial consequences. The cost of recovery often runs into millions of dollars, a staggering figure that only tells part of the story. Beyond the immediate financial impact lies an equally daunting challenge: repairing the company's damaged reputation. This intangible yet critical asset, once tarnished, can have long-lasting effects on customer trust, market position, and future business prospects.

For example: In a case where a business is hit with Ransomware (a virus that locks victims' computers, and the attacker threatens to publish the victim's personal data or permanently block access to it unless a ransom is paid off), and they decide not to pay the ransom, and the attacker actually releases the data of let's just say 50,000 users. The business will be forced to pay for identity protection for the victims.

So if it costs $35 per person per month, let's do the costs for 5,000 victims for just one year.

$$1 \text{ victim} = \$35 \text{ per month}$$

$$1 \text{ victim for a year} = \$35 \times 12 = \$420$$

Now if you have 5,000 victims = 50,000 × 420 = **$21,000,000** for the year!!

Can you imagine a business spending that much for just Identity protection on victims? Most businesses will not survive this type of attack.

As such, protecting these revenue-generating systems and data becomes a top priority, highlighting the essential role that cybersecurity professionals play in today's business environment. In the upcoming chapters, we will delve deeper into how cybersecurity intersects with business operations and the strategies to land your first job in this field.

1.5 Conclusion

With a basic understanding of what businesses are and why they exist, we've taken the first step on our cybersecurity journey. In the next chapter, we'll delve into the role of information technology (IT) in businesses and how it has changed the way they operate.[1]

Knowledge Check

Question 1: What is the main reason businesses exist?
Answer: To make a profit.

Question 2: How do businesses generate revenue?
Answer: By selling goods or services.

Question 3: What is the difference between revenue and profit?
Answer: Revenue is total income before expenses. Profit is revenue minus expenses.

Note

1 Braue, D. (2021, June 2). Global Ransomware Damage Costs Predicted To Exceed $265 Billion By 2031. *Cybersecurity Ventures*. https://cybersecurityventures.com/ransomware-report-2021/

2
The Role of IT in Business

2.1 Introduction

Now that we have an understanding of what a business is and how it generates revenue, let's turn our attention to a critical component that has revolutionized modern businesses – information technology (IT). Just as a deep understanding of businesses is crucial, understanding IT's role and impact on businesses is equally important for aspiring cybersecurity professionals.

2.2 What is Information Technology?

IT involves the use of computers, storage, networking, and other physical devices, infrastructure, and processes to create, process, store, secure, and exchange all forms of electronic data. Businesses use IT to improve their efficiency, create innovative products, reach a wider customer base, and compete on a global scale.

The Strategic Role of IT in Business

Driving Business Processes: IT streamlines business processes, enhancing efficiency and productivity. Automated workflows and data analytics empower businesses to make informed decisions swiftly.

Enabling Globalization: IT connects businesses with international markets through e-commerce platforms, allowing them to expand their reach and operate globally with relative ease.

Facilitating Innovation: Digital platforms enable businesses to innovate their products and services. Cloud computing, AI, and IoT are just a few technologies spurring new business models and opportunities.

DOI: 10.1201/9781003501589-3

2.3 Emerging IT Trends and Their Business Implications

- **Artificial Intelligence and Machine Learning:** AI and ML are transforming customer service, operational processes, and decision-making, allowing businesses to offer personalized experiences and automate complex tasks.
- **Blockchain:** Beyond cryptocurrencies, blockchain technology offers secure and transparent ways to conduct business transactions and manage contracts, significantly impacting supply chain management and financial services.
- **Cloud Computing:** The shift to cloud services offers businesses flexibility, scalability, and cost efficiency, enabling them to adapt to market demands and innovations rapidly.

2.4 IT's Role in Competitive Advantage

- **Data Analytics:** Leveraging big data and analytics, businesses can gain insights into customer behavior, market trends, and operational efficiency, guiding strategic decisions that enhance competitiveness.
- **Cybersecurity as a Competitive Edge:** In an era where data breaches can tarnish a company's reputation, robust cybersecurity measures not only protect the business but also strengthen customer trust and loyalty.
- **Agility and Responsiveness:** IT enables businesses to be more agile and responsive to market changes, ensuring they can capitalize on opportunities and mitigate risks effectively.

2.5 Integrating Cybersecurity into Business Strategy

- **Risk Management:** Cybersecurity is integral to risk management strategies, protecting business assets from cyber threats and ensuring business continuity.
- **Regulatory Compliance:** Navigating the complex landscape of data protection laws requires a strategic approach to IT governance and compliance.

- **Building a Security-Conscious Culture:** Training and awareness programs are essential to foster a culture where every employee understands their role in maintaining cybersecurity.

2.6 Conclusion

The role of IT in business extends far beyond managing infrastructure and support services. It is a key driver of innovation, competitive advantage, and strategic growth. As businesses continue to evolve in the digital age, integrating IT and cybersecurity into the core business strategy will be paramount for success.

In the next chapter, we will explore how cyber threats have emerged alongside technological advancements and the implications for businesses.

Knowledge Check

Question 1: How has IT impacted modern businesses?
Answer: Made them more efficient, interconnected, and globally competitive.

Question 2: Name two ways IT can help businesses increase revenue.
Answer: E-commerce platforms and cloud-based subscription services.

Question 3: What are two cybersecurity risks introduced by reliance on IT?
Answer: Data breaches and disruption of operations from cyberattacks.

3
The Pillars of IT and Cybersecurity

3.1 Introduction

With the proliferation of IT in businesses, cyber threats have risen in tandem. These threats pose significant risks to businesses of all sizes. However, you cannot protect what you can't see. To be able to go out into the world as true Cyber professional you have to understand the layering behind that job posting and how that came into being.

3.2 The Evolution of IT and Cyber Threats

IT's growth can be broadly categorized into three pillars - Hardware, Networking, and Databases.

- **Hardware:** In the early stages, businesses heavily relied on hardware like computers, servers, and storage devices. These were susceptible to physical damages, malfunctions, and wear and tear, requiring regular maintenance and replacements.
- **Networking:** As technology advanced, businesses started connecting their hardware to form networks. In essence, Networking is where two computers/devices can talk to each other and share information and resources seamlessly.
- **Databases:** Once that information was shared between devices, there had to be a way to store that information. The advent of databases allowed businesses to store, manage, and analyze vast amounts of data efficiently.

These 3 pillars were supposed to make life and doing business much easier, However, they also exposed businesses to new threats. Unauthorized users could gain access to these networks, disrupt their functioning, or steal valuable data. Databases became lucrative targets for cybercriminals seeking to steal or manipulate data.

DOI: 10.1201/9781003501589-4

3.3 Layering Security on IT

As IT evolved and became increasingly integral to businesses, so did the need to protect its components. This need led to the development and implementation of cybersecurity measures, layered on top of the IT pillars.

- **Hardware Security:** This involves physical security measures to protect hardware from theft, damage, or tampering. Additionally, it includes safeguards like access control, disk encryption, and secure hardware disposal procedures.
- **Network Security:** This focuses on protecting the integrity and usability of network and data. It includes protective measures like intrusion detection systems (IDS), firewalls, and secure network architectures to prevent unauthorized access and data breaches.
- **Database Security:** This includes encryption, backup, access controls, and other measures to protect databases from unauthorized access, data breaches, and data loss.

Now, let's look at some key terms you're going to see everywhere in the cyber world.

- **Threat:** Think of a threat as potential bad weather or a burglar that might want to break into your house.

  Example: A thunderstorm approaching your neighborhood or someone lurking around homes, looking for an easy break-in.
- **Vulnerability:** This is like having a broken window latch or a door that doesn't lock properly in your house. It's a weak spot that a burglar (threat) could exploit.

  Example: Your back door has a faulty lock that doesn't secure properly.
- **Risk:** The combination of a threat and a vulnerability. For instance, the risk is high if there's a burglar nearby and you have that faulty back door. It means there's a good chance the burglar could use that door to enter your house.

  Example: The likelihood of a break-in because of that faulty back door, especially when you know there have been burglaries in your area.

- **Risk Transfer:** Imagine you decide to get insurance for theft. Even if a burglar breaks in and steals something, the insurance company will cover your loss. You're transferring the financial burden of the risk to someone else.
  Example: Purchasing home insurance that covers theft.

3.4 The Implications of Cyber Threats

Cyber threats pose significant risks to businesses. These risks can result in financial loss, reputational damage, legal liabilities, and even loss of business operations. Small businesses may lack robust security measures, making them particularly vulnerable. Meanwhile, larger businesses may be targeted for their wealth of valuable data.

Imagine you have a precious item, let's say a rare jewel. You'd want to protect it from theft, damage, or loss, right? So, besides keeping it safe, you might also get insurance for it. If, unfortunately, it gets stolen or damaged, the insurance can help cover the cost of replacing or repairing it.

Now, let's think of your business's digital assets – like your customer data, your website, or even your online brand reputation – as that rare jewel. In today's digital world, there are many 'digital thieves' and various mishaps that can harm these assets. These are what we refer to as cyber threats.

To address such threats, businesses might choose to transfer their risks (**Risk Transfer**). As previously mentioned, one approach to this is by obtaining **Cyber Insurance** to safeguard their digital assets.

So, what is Cyber Insurance?

Cyber insurance is like a safety net for businesses in the digital age. It's a specialized insurance plan designed to protect businesses against potential losses from cyber-related events. These could be events like data breaches, cyberattacks, or even instances of employee negligence leading to data exposure.

Why might businesses need it?

Financial Coverage: Dealing with a cyber incident can be expensive. From hiring experts to stop a cyberattack, notifying affected customers, to potential legal costs – the bills pile up. Cyber insurance helps cover these costs.

Expertise Access: Cyber insurance often provides access to cybersecurity professionals who can help manage and mitigate a breach when it occurs.

Peace of Mind: Just as with any insurance, it provides businesses the peace of mind knowing they have a backup plan in case things go south.

However, it's essential to understand that while cyber insurance can provide a safety net, it's not a replacement for robust cybersecurity practices. It's like having health insurance but still eating well and exercising to stay healthy. Both the preventive measures (cybersecurity practices) and the safety net (cyber insurance) are crucial in today's digital business landscape

3.5 Cyber Security Career Paths

Now that you've got a grasp on the basics of cybersecurity, let's dive into the various career paths you can explore. Take your time going through this list—it's designed to help you find the path that resonates with you. The beauty of cybersecurity is that it truly is for everyone. No matter your background, it can actually enhance your abilities as a cybersecurity expert. For instance, if you're a teacher, you might wonder, "How can I transfer my skills into cybersecurity?" Well, get ready, because I'm about to show you how!

1. Cybersecurity Awareness and Training:
    - What they do: Develop and deliver training programs to employees about the latest cybersecurity threats and the best practices to counteract them.
    - Why teachers fit: Teachers have the skills to create curriculum, understand different learning styles, and deliver information in a way that is engaging and memorable.
2. Security Documentation and Policy Development:
    - What they do: Write and update the policies, procedures, and guidelines that govern how an organization and its employees should approach security.
    - Why teachers fit: Teachers are adept at creating structured, clear, and actionable lesson plans, which can translate into drafting effective security policies and procedures.

3. Security Evangelist or Advocate:
   - What they do: Promote security awareness both inside and outside of an organization. They might give talks, write articles, or engage in community outreach.
   - Why teachers fit: Teachers are natural communicators and can convey complex ideas in a way that is accessible to a diverse audience.
4. Cybersecurity Sales or Product Training:
   - What they do: In companies that sell cybersecurity products or services, trainers are needed to educate potential clients or internal teams about the product's benefits and usage.
   - Why teachers fit: Teachers can break down complicated topics (like a new software) into understandable chunks and deliver them effectively.
5. User Experience (UX) in Security Products:
   - What they do: Work with design teams to make sure that security tools and software are user-friendly.
   - Why teachers fit: Teachers understand how users (in this case, students) think and can anticipate areas of confusion or difficulty.
6. Roles in Governance, Risk Management, and Compliance (GRC):
   - What they do: Oversee the strategy and processes to manage cybersecurity risks, ensure compliance with laws and regulations, and align with business objectives.
   - Why teachers fit: They can effectively communicate policies, ensure that processes are followed, and identify potential areas of risk or non-compliance.

Wow'ed yet? Go over this list and pick 3 paths that speak to you or your interests.

JOB TITLE	DESCRIPTION
Security Analyst	Monitor and defend systems from unauthorized activity.
Network Security Engineer	Protect the company's network and system from threats.
Cybersecurity Analyst	Protect an organization's data and network, assess vulnerabilities, and handle incidents.

(Continued)

JOB TITLE	DESCRIPTION
Cryptographer	Develop algorithms, ciphers, and security systems to encrypt sensitive information.
Forensic Expert	Identify, investigate and prevent cyber crimes by analyzing hard drives, network logs etc.
Security Consultant	Provide security advice based on risk assessments and knowledge of current threats and solutions.
Ethical Hacker	Test systems and networks for vulnerabilities that malicious hackers could exploit.
Cybersecurity Manager/ Administrator	Develop and implement security policies and procedures to protect an organization's computer networks and systems.
Chief Information Security Officer (CISO)	Executive responsible for an organization's information and data security strategy.
Cybersecurity Engineer	Design and implement secure network solutions to defend against advanced cyberattacks.
Penetration Tester	Conduct simulated attacks to identify vulnerabilities before actual hackers find them.
Security Architect	Design, create and monitor the deployment of an organization's network and computer security.
Information Security Analyst	Plan and carry out security measures to protect a company's computer networks and systems.
IT Security Consultant	Advise clients on their security policies, identify risks and propose solutions.
Security Systems Administrator	Manage and administer an organization's security solutions.
Incident Responder	Handle security incidents, analyze them and recommend improvements to avoid future incidents.
Cybersecurity Sales Engineer	Provide pre-sales support by understanding customer requirements and designing solutions.
Security Auditor	Check systems for vulnerabilities, reporting findings to management.
Vulnerability Assessor	Identify and quantify the security vulnerabilities in an organization's systems.
Cybersecurity Software Developer	Develop security software and integrate security into software during the design and development process.
Security Code Auditor	Review the code to find any security flaws and vulnerabilities.
Cyber Crime Investigator	Investigate crimes on the internet and digital evidence of crimes.
Information Risk Auditor	Evaluate, test, and report on the effectiveness of security controls.
Cyber Insurance Specialist	Assess a company's risk profile and recommend appropriate cyber insurance policies.
Cybersecurity Project Manager	Manage and coordinate cybersecurity projects within an organization.

(*Continued*)

JOB TITLE	DESCRIPTION
Security Operations Center (SOC) Analyst	Monitor and analyze activity on networks, servers, endpoints, databases, applications, websites, and other systems for any security breaches.
Governance, Risk Management & Compliance (GRC) Specialist	Help ensure the organization meets its security, privacy, and risk compliance obligations.
Disaster Recovery and Business Continuity Analyst	Plan, implement and maintain controls that help the organization in the event of an incident or crisis.

Remember that the specific tasks and responsibilities of these roles can vary greatly depending on the size, type, and security needs of the organization.

3.6 Conclusion

In essence, the growth and evolution of IT in businesses brought along the emergence of cyber threats. As a cybersecurity professional, your role is to understand these threats and layer security measures on top of the IT infrastructure to protect businesses' valuable assets. The subsequent chapters will delve into the specifics of cybersecurity, including its key areas and concepts, equipping you with the knowledge to fulfill this critical role.

Knowledge Check

Question 1: As dependence on IT grew, what emerged in parallel?
Answer: Cyber threats

Question 2: What are the three pillars of IT evolution?
Answer: Hardware, Networking, Databases

Question 3: Name one implication of cyber threats for businesses.
Answer: Financial loss, reputational damage, disruption of operations.

4

Simplifying Cybersecurity

Key Areas and Concepts

4.1 Introduction

Stepping into the sphere of cybersecurity might seem intimidating with all its terminologies and concepts. Let's try to simplify it, so it becomes as familiar as building a house. Just like constructing a safe and secure home involves planning, understanding materials, and following building codes, cybersecurity relies on principles, concepts, and guidelines. Let's make you the architect of your cybersecurity knowledge!

4.2 The CIA Triad: A Blueprint of Security

When you're constructing a house, there are three aspects that you always consider: privacy, structure, and accessibility. These are remarkably similar to the CIA Triad: Confidentiality, Integrity, and Availability in cybersecurity.

- **Confidentiality (Privacy):** Just like high walls, sturdy doors, and windows with curtains maintain your home's privacy, confidentiality ensures unauthorized individuals cannot access your information. This can be achieved with user IDs, strong passwords, and data encryption methods, acting as your cybersecurity walls and locks.
- **Integrity (Structure):** You wouldn't want anyone to manipulate the structure of your house without your knowledge. Integrity in cybersecurity does exactly this – it guarantees that data remains untouched and unchanged unless authorized. Techniques like version control and checksums are the 'construction supervisors,' making sure no unauthorized changes are made.

DOI: 10.1201/9781003501589-5

- **Availability (Accessibility):** A house is only useful if you can enter when needed. Similarly, in cybersecurity, availability ensures authorized users have access to data and resources whenever required. This is like having a spare key; mechanisms like data redundancy and disaster recovery are your spare keys to the information, making sure it's available when you need it.

4.3 Understanding IAAA: The Construction Codes

When you build a house, specific rules must be followed. Similarly, cybersecurity follows the IAAA protocol: Identification, Authentication, Authorization, and Accountability.

- **Identification:** Like placing a nameplate on your house, users declare their identity to a system, typically with a username or email.
- **Authentication:** This is the process of verifying the declared identity. It's like using a key to prove that you are the house's owner. This could be a password, fingerprint, or even facial recognition.
- **Authorization:** Once authenticated, you're allowed to perform certain actions, like opening the door or window. In a system, certain privileges and rights are granted to the user.
- **Accountability:** This is like a security camera that records all actions in your house. In cybersecurity, all user activities are logged and can be audited.

4.4 The Importance of Cybersecurity Frameworks

Just as a sturdy house needs a solid blueprint, robust cybersecurity requires a comprehensive framework. In the context of building a house, a blueprint outlines the structure, detailing where the foundation should be, how the walls should be assembled, and how to construct the roof. Similarly, a cybersecurity framework offers a detailed guide to secure an organization's information technology structure. It identifies what measures should be put in place, how these should be managed, and how to respond if a security incident occurs.

Different companies align with different cybersecurity frameworks, based on their specific needs, industry requirements, regulatory environment, and geographical location. Let's take a brief look at some of the most popular frameworks and their significance:

- 🚀 **National Institute of Standards and Technology (NIST):** Think of NIST as the Grandmaster of Cyber Lore. Their framework serves as a trusty map guiding organizations through the treacherous lands of cyber risks. Embracing risk and flexing with all sorts of tribes, big and small, NIST is like the Swiss Army knife of cyber guidelines. From health records to banking details, NIST watches over all.
- 🛡 **International Organization for Standardization (ISO):** The Knights of ISO 27001 guard the kingdom's most treasured secrets with a systematic finesse. When it comes to fortifying your realm's sensitive info, the ISO brigades got your back. Manufacturing and technology firms love this framework.
- 🔍 **Control Objectives for Information and Related Technologies (COBIT):** In the grand citadel where tech wizards and business lords coexist, COBIT weaves magic to bridge their worlds. Especially cherished by enterprises navigating the intersection of technology and business, it ensures the gears of commerce churn smoothly.
- 🎭 **Committee of Sponsoring Organizations of the Treadway Commission (COSO):** Donning masks of financial, operational, and compliance warriors, the COSO clan is known for their prowess in enterprise risk battles, ensuring no IT dragon goes unchecked.
- 🌀 **NIST Cybersecurity Framework (CSF):** The NIST CSF is like the chosen one, a beacon of hope in the vast expanse of the cyber universe. Crafted to shine a guiding light for the private sector, this starry framework charts the course for identifying treacherous cyber risks and vanquishing them.

These frameworks aren't mutually exclusive and can be used in conjunction to provide a comprehensive approach to cybersecurity. For instance, an organization might use NIST for its overall cybersecurity strategy, ISO 27001 for its information security management, and COBIT for its IT governance. The choice of the framework will depend on the organization's unique circumstances, including its size, industry, and the nature of its data. In the next sections, we'll delve deeper into each of these frameworks and discuss how they can be used in practice.

4.5 NIST Cybersecurity Framework: The Construction Manual

Building a house requires following a detailed plan or manual. The NIST has created a similar guide called the cybersecurity framework, which outlines steps for managing cybersecurity risks.

- **Identify:** Understand the 'blueprints' of your business – the resources, and potential cybersecurity risks.
- **Protect:** Establish safeguards, much like installing security systems in your house to ensure safety.
- **Detect:** Keep an eye out for any suspicious activities or events, like a security guard patrolling around your house.
- **Respond:** Have a plan to take immediate action when a security incident occurs, much like calling the police when a break-in happens.
- **Recover:** Develop strategies to restore any capabilities or services impaired due to a cybersecurity incident, much like restoring your house after a natural disaster.

4.6 Risk Management: The Safety Inspection

Finally, building a safe house involves assessing and managing risks. In cybersecurity, Risk Management follows a similar principle. It involves identifying potential risks, assessing their potential impact, and then deciding on a course of action. For instance, if your house is in a flood-prone area, you'd assess the risk, possibly elevate the house or have water barriers, and have an evacuation plan ready. Similarly, in cybersecurity, you might identify a risk like potential data breach, assess its impact, and then decide to enhance your firewall or encryption techniques to mitigate this risk.

Risk Mitigation: This is like fixing the faulty lock on your back door or setting up a security camera. You're taking steps to reduce the chance of a burglar entering your home.

Example: Installing a new lock or getting a guard dog.

4.6.1 Security Tools and Descriptions

Building a secure 'house' in the context of cybersecurity isn't so different from constructing a physical house. When building a house, you start with a solid foundation, erect strong walls, install secure doors and windows, set up a reliable alarm system, and ensure there's a contingency plan for emergencies. Every element of this house is essential for its overall protection, much like every tool and function in cybersecurity is critical for the overall protection of a business.

Now let's explore the elements of our cybersecurity 'house' and the corresponding tools and functions that ensure its security:

- **Foundation (EGRC):** Enterprise Governance, Risk, and Compliance (EGRC) forms the bedrock of the cyber 'house'. It underpins all activities, much like a foundation supports all parts of a physical house. **ServiceNow's GRC** module is an example of an EGRC application that manages governance, risk, and compliance with regulations.
- **Structural Assessment (Vulnerability Scanning):** Just as a building inspector assesses the structure of a house, vulnerability scanning tools like **Nessus** and **HCL AppScan** assess systems for any weaknesses that could be exploited by attackers.
- **Walls (Application Security):** Application security tools like **Veracode** and **HCL AppScan** form the walls of the cyber 'house', identifying, fixing, and preventing security vulnerabilities.
- **Doors and Windows (Identity Management Tools):** Much like doors and windows control who enters or exits a house, Identity Management Tools such as **Okta** and **Microsoft Azure Active Directory** control access to systems and data.
- **Alarm System (Threat Hunting and SIEM):** Threat Hunting tools like **CrowdStrike's Falcon** and SIEM (Security Information and Event Management) systems like **LogRhythm** and **Splunk** work as the alarm system, providing real-time analysis and alerts for potential threats.
- **Barricades (DDoS Protection):** DDoS protection tools like **Cloudflare** and **Amazon AWS Shield** act as barricades to your cyber 'house', guarding against overwhelming traffic that can disrupt your services.

- **Safe (Data Loss Prevention – DLP):** DLP tools like **Forcepoint DLP** and **Symantec (Broadcom) DLP** function like a safe, preventing potential data leaks by detecting and blocking sensitive data.
- **Emergency Plan (Threat Intelligence):** Just as an emergency plan prepares homeowners for unexpected disasters, Threat Intelligence tools like **Recorded Future** and **FireEye** equip businesses with information about emerging or existing threats, enabling them to respond effectively.
- **Maintenance and Upkeep (Security Education Content Creation):** Regular maintenance is essential to keep a house safe and up-to-date. In cybersecurity, tools like **Articulate 360** serve a similar role by creating training modules to educate employees on cybersecurity best practices.

This list is by no means exhaustive and cybersecurity is a continually evolving field. As you embark on your cybersecurity journey, you will encounter and learn about many more tools that specialize in various areas. Remember, the best tool depends on the specific needs and context of the business. Just like the perfect house, the perfect cybersecurity setup is different for everyone!

4.7 Conclusion

With these analogies, we hope the world of cybersecurity appears less daunting and more familiar to you. Like constructing a house, it involves planning, understanding materials (concepts), and following building codes (guidelines). As we venture further, we'll help you master the skills of cybersecurity, akin to becoming a master architect of your own secure house!

Knowledge Check

Question 1: What are the three aspects of the CIA triad?
Answer: Confidentiality, Integrity, Availability.

Question 2: What do the four steps of IAAA stand for?
Answer: Identification, Authentication, Authorization, Accountability.

Question 3: Name one popular cybersecurity framework.
Answer: NIST, ISO 27001, CIS, COBIT.

5

Preparing for the Job Hunt

What You Need to Know

Embarking on your job hunt in the world of cybersecurity may seem like an intimidating journey, especially if you are just starting out. However, let this chapter serve as your compass, guiding you through the process of job interviews in the cybersecurity field. We'll demystify the complexities of scenario-based and technical questions and help you develop the art of asking insightful questions.

5.1 Understanding the Interview Process

The job interview process in cybersecurity, like many fields, typically consists of three main stages:

- **Preliminary Screening:** Recruiters or HR representatives conduct this initial stage to verify the information on your resume and assess whether you align with the company's culture and the role's salary and location requirements.
- **Technical Interviews:** If you pass the preliminary screening, you'll interview with potential team members. These interviews focus on the technical skills required for the role. For a Security Operations Center (SOC) position, for instance, you might be asked about TCP/IP, DNS, routing, the Cyber Kill Chain, and encryption.
- **Final Interview:** This is usually with the hiring manager, who makes the final hiring decision. They assess your fit with the team and the alignment of your skills with the role's needs.

DOI: 10.1201/9781003501589-6

5.1.1 Job descriptions

Your chances of landing any job in cyber security increase by as much as 50% if you understand the job description really well. In the example below, we're going to look at a Sample Security Analyst job description and look at how the sections have been broken down. Consider what you're supposed to focus on for each line.

Job Title: Cyber Security Analyst (Entry Level)

5.1.2 Company Overview

XYZ Tech Solutions is a leading digital services company with a focus on ensuring data security for its clients worldwide.

Comment: The company overview gives you insight into the company's primary domain. Tailor your application by researching more about the company and integrating it into your cover letter or discussions during the interview.

5.1.3 Responsibilities

Monitor and respond to alerts from security tools; investigate potential security breaches and incidents.

Comment: This line indicates a requirement for basic troubleshooting and analytical skills. Highlight any relevant coursework or projects that involved problem-solving.

Perform vulnerability assessments on company systems to identify potential weaknesses and recommend remediation measures.

Comment: Focus on any experience or understanding you have about vulnerability scanning tools or practices, even if it was in an academic setting.

Assist with the implementation and maintenance of security tools and solutions.

Comment: This shows they're looking for someone who's hands-on. Be ready to discuss any practical experience with security tools or coursework that had hands-on labs.

Participate in cybersecurity awareness training sessions for employees, helping them understand and follow best practices.

Comment: Emphasize any teaching, mentoring, or presentation experiences. This implies good communication skills and the ability to simplify complex topics.

Stay updated with the latest cybersecurity threats and trends to ensure company systems remain secure.

Comment: Demonstrates the importance of continuous learning. Mention any cybersecurity blogs, podcasts, or news sources you follow.

5.1.4 Qualifications

Bachelor's degree in information technology, computer science, or related field.

Comment: If your degree isn't directly related but you've done relevant coursework or projects, highlight those in your application/ resume.

Basic knowledge of common cybersecurity tools (e.g., firewalls, intrusion detection systems) and principles.

Comment: Familiarize yourself with popular cybersecurity tools in the industry. Bring up any hands-on experience, even if self-taught.

Strong analytical and problem-solving skills.

Comment: Be ready to provide examples of challenges you've faced and how you've overcome them.

5.1.5 Desired but not essential

Certifications such as Security+, CEH, or any other entry-level security certification.

Comment: If you have a certification, great! If not, consider mentioning your plan to pursue one in the near future.

Experience with scripting languages (Python, Bash).

Comment: Even basic scripting can be a strong point. If you've automated any task using a script, be sure to mention it.

Benefits

Competitive salary package

Health and dental insurance

Ongoing professional development opportunities

Comment: Consider these benefits when negotiating. If they don't match your expectations, there might be room for negotiation, especially if you bring something valuable to the table.

5.1.6 Application Process

Please send your updated resume, a cover letter detailing your relevant experience, and any supporting documents to [email@email.com].

Comment: Tailor your cover letter to the job description. Address the key responsibilities and qualifications mentioned, drawing parallels to your experiences.

Applicants will be considered for interviews based on the criteria listed in the job description. We thank all applicants for their interest, but only those selected for an interview will be contacted.

Comment: Be patient after applying. Depending on the number of applicants, it might take time for the company to review all applications and respond.

5.2 Interview Preparation

Preparation is key for successful interviews. Research the company, understand its mission and culture, and familiarize yourself with the role you're applying for. Position yourself as an ideal candidate by aligning your skills and experiences with the job description.

5.2.1 The RUDP Method: Your Secret Weapon for Cybersecurity Interview Success

Before we dive into the nitty-gritty, let's get a bird's-eye view of the RUDP method. This framework is your roadmap to acing those cybersecurity interviews. Here's what it stands for:

R - Read: Dive deep into the company and industry. U - Understand: Decode what the job really requires. D - Decipher: Break down the technical requirements and buzzwords. P - Prepare: Bring it all together for interview day.

Think of RUDP as your personal interview prep checklist. It's designed to transform you from just another candidate to the one they can't wait to hire. Now, let's break it down and see how this method can supercharge your interview game.

R - Read (Like Your Career Depends On It)

First up, channel your inner Sherlock Holmes. We're not just skimming here; we're doing a deep dive into:

- The company's story (their website is your new favorite bedtime reading)
- Industry reports (yeah, they can be dry, but they're gold mines of info)
- Breaking news (mergers, hacks, product launches – if it's happening, you need to know)

Why go all detective mode? Because when you drop these knowledge bombs in your interview, you're basically saying, "I'm not just here for a job; I'm here to be part of your story." That's the kind of enthusiasm that gets you hired!

Pro Tip: Create a mini-dossier on the company. It's like your personal cheat sheet for the interview.

U - Understand (Decode the Job Description Like a Pro Hacker)

Now, let's crack that job description wide open. We're looking for:

- Must-have skills (these are your non-negotiables)
- Nice-to-have skills (your bonus points)
- The real deal behind the corporate speak (Are they after a future CISO or a technical wizard?)

Once you've got this figured out, it's story time! Start lining up your experiences with what they're after. Got a tale about leading a project? Polish it up. Solved a tricky security puzzle? Get ready to share it.

Remember: Every line in that job description is a chance to say, "That's me you're looking for!"

D - Detect / Decipher (Time to Geek Out and Level Up)

Grab your highlighter and go to town on that job description. Every tech term, every buzzword – highlight it!

- EDR? SIEM? Zero Trust? Circle 'em all!
- For each term: If you know it, great! Prep an example. If it's new, congrats! You've got homework.

Real talk: I once saw "Experience with threat hunting" in a job desc. Did I panic? Nope! I dug into some online courses, practiced with some open-source tools, and boom – I had something solid to talk about in the interview.

Bonus Challenge: Create a mind map linking these keywords to your experiences or learning plans. It's like building your personal cybersecurity skill tree!

P - Prepare (Because Winging It Is So Last Season)

This is where you bring it all together:

1. LinkedIn stalking (professionally, of course) is now your part-time job. Find out who's interviewing you.
2. Did your interviewer write a blog post or speak at a conference? That's your required reading/watching.
3. Craft questions that make them think, "Wow, this candidate really gets us!"

Example questions to knock their socks off:

- "I noticed your recent implementation of a zero-trust architecture. How has this impacted your incident response procedures?"
- "Your CEO mentioned prioritizing AI in cybersecurity in her last keynote. How is this shaping your team's strategy?"

Remember, by doing this, you're not just another applicant. You're the candidate who's already thinking like part of the team.

Bonus Round: Interview Day Power-Ups

1. Create checklists for each RUDP stage. Check them off like a boss!
2. Rope in a friend for mock interviews. The more you practice, the more you'll shine.
3. After each interview, do a personal debrief. What worked? What didn't? Continuous improvement is the name of the game.

Remember, rockstar, this method isn't just about acing the interview (though you totally will). It's about finding your perfect fit in the cybersecurity universe. It's about walking into that room not just as a candidate, but as the solution to their problems.

5.2.2 Interview Questions and Answers

During the interview, expect a mix of scenario-based and technical questions.

- **Scenario-based questions** assess your problem-solving skills and how you handle specific situations. For example, you might be asked:

  Question: "Imagine you detect a potential security breach. What steps do you take?"

  A good answer might be: *"First, I would follow the company's established incident response plan, which likely includes isolating the affected systems to prevent further damage. Next, I would gather as much information as possible about the breach – what systems are affected, how it was discovered, etc. – to aid in the investigation. Then, I'd report the incident to my supervisor and the appropriate team, such as IT or legal, depending on the company's protocol."*
- **Technical questions** evaluate your technical knowledge and skills. For instance:

  Question: "Can you explain what DNS is and why it's important?"

  A possible answer could be: *"DNS, or Domain Name System, is essentially the phonebook of the internet. It translates human-friendly domain names like 'www.example.com' into the IP addresses that computers use to identify each other. Without DNS, we would have to memorize complex IP addresses to navigate the internet. In terms of security, DNS is crucial because attackers often exploit it to direct users to malicious websites."*

Remember, interviews are also an opportunity for you to ask questions. Asking about the company's cybersecurity strategies or about the team you'll be joining shows you're proactive and interested.

5.3 Sample Questions To Ask the Hiring Manager

Here are some good questions an entry-level cybersecurity candidate could ask during a job interview:

- What are the day-to-day responsibilities of this role? This helps the candidate understand the specific duties they would be performing.

- What technologies and tools does your cybersecurity team use on a regular basis? This gives insight into what skills and experience they will gain on the job.
- How does the cybersecurity team interact with other departments like IT, legal, and executives? This sheds light on how collaborative the role is.
- What training and professional development opportunities are available for cybersecurity team members? This shows if there is room for growth.
- What are some of the biggest cybersecurity challenges facing your company right now? This lets the candidate know what types of projects they may be involved in.
- Does your company follow any cybersecurity frameworks like NIST or ISO? This demonstrates the candidate's knowledge of industry standards.
- What qualities make someone successful on your cybersecurity team? This gives the candidate an idea of the soft skills that are valued.
- Is there flexibility in the work schedule/remote work opportunities? This allows the candidate to understand the work arrangements.
- What is the typical career path for someone starting in an entry-level cybersecurity role? This gives visibility into advancement opportunities.
- How is the cybersecurity team structured? This helps the candidate see where they would fit into the organizational chart.
- What is the budget and headcount for the cybersecurity team? This demonstrates the candidate's interest in the group's resourcing.

5.4 Conclusion

Succeeding in cybersecurity job interviews requires thorough preparation and practice. Research the company and role to understand how to position yourself as an ideal candidate. Expect and rehearse responses for both scenario-based questions that assess problem-solving skills and technical questions that evaluate your knowledge. Come armed with insightful questions that demonstrate curiosity and strategic thinking about the organization's security priorities and culture. Show enthusiasm and avoid negativity when discussing your background. With technical expertise finely tuned and excellent communication skills, you can tackle any cybersecurity interview question thoughtfully and make a compelling case for why you are the best fit for the role. Remember that interviews are a two-way street – you are also assessing if the company is the right environment for you to grow in your cybersecurity career.

Knowledge Check

Question 1: What are the three main stages of a cybersecurity job interview process?
Answer: Screening, Technical Interview, Final Interview.

Question 2: What two types of interview questions should you expect and prepare for?
Answer: Scenario-based and Technical.

Question 3: True or False: You should only answer interview questions, not ask your own.
Answer: False.

6 MASTERING YOUR RESUME

Your resume is your handshake before you walk into the room. It is your silent ambassador, speaking volumes about you even before you utter your first word. Understanding its importance, you must craft it meticulously, ensuring it embodies your skills, knowledge, and experience.

Active vs Passive Voice: The active voice is direct and impactful, adding a sense of dynamism and responsibility to your achievements. "Managed a team of 10 analysts" sounds more assertive and powerful than "A team of 10 analysts was managed." Wherever possible, use the active voice to make your statements more compelling.

Including Metrics: Quantifying your achievements lends credibility to your claims. For instance, instead of saying "Improved system security," say "Enhanced system security by 30% by implementing new protocols." These specifics give a tangible dimension to your contributions.

Structuring Your Resume: The way you organize your resume can enhance its impact. It's best to start with a summary that encapsulates your skills, experiences, and career aspirations, followed by sections for your education, work experience, technical skills, certifications, and other relevant accomplishments.

Tailoring Your Resume: Your Secret Weapon for Landing Interviews.

Listen up, cyber warriors! We're about to unlock a game-changing strategy that could double your chances of scoring that coveted interview. I'm talking about tailoring your resume for each job application. Yeah, I know what you're thinking - "But that sounds like a lot of work!" Well, buckle up, because I've got a trick up my sleeve that's going to make this process a breeze.

Why Tailor Your Resume?

Sure, having a one-size-fits-all resume is convenient, but here's the deal: when you customize your resume for each job, you're not just

 DOI: 10.1201/9781003501589-7

ticking boxes - you're showing that you're the missing puzzle piece they've been searching for. By aligning your skills and experiences with the specific requirements of the job, you're essentially saying, "Hey, I'm not just qualified - I'm tailor-made for this role!"

And get this: a personalized resume can double your chances of landing an interview. That's right, DOUBLE! In the competitive world of cybersecurity, that's an advantage you can't afford to pass up.

Enter the AI Resume Tailor – (https://chatgpt.com/g/g-mcmbgihk9-cyber-resume-tailor)

Now, I know what you're thinking. "Sounds great, but who has time to rewrite their resume for every single application?" That's where my secret weapon comes in - a free AI engine that I've curated just for you. This isn't just any AI; it's your personal resume tailoring assistant.

Here's how to use this cyber-age magic:

1. Start with your current resume - you know, the one you've been sending everywhere.
2. Find a job posting that makes your cyber-heart skip a beat.
3. Fire up the AI engine and give it this simple prompt:
   "Read my current resume. Now compare it to the attached job description (go ahead and paste the job description here). Now, tailor my resume for this job description."
4. Sit back and watch the AI work its magic.
5. In just a few moments, voilà! You'll have a shiny new resume, perfectly tailored to that dream job.
6. Download or copy-paste your new resume, and you're ready to apply!

It's that simple. No more stressing over which skills to highlight or which experiences to emphasize. The AI does the heavy lifting, leaving you more time to practice your interview skills (hint: check out the RUDP method we discussed earlier).

Remember, in the world of cybersecurity, staying ahead of the curve is key. By leveraging AI to tailor your resume, you're not just showcasing your skills - you're demonstrating your ability to use cutting-edge tools to solve problems. And isn't that exactly what a great cybersecurity professional does?

So, what are you waiting for? Give it a try with your next application. Your future self (you know, the one with the awesome new cyber job) will thank you!

6.1 Certifications

It seems there's this comedic misconception floating around, where folks picture getting cybersecurity certifications as a vending machine – pop in certification and out-tumbles a million dollars. Like believing you can become a master chef by merely boiling an egg. Well folks, spoiler alert: reality doesn't work on sitcom logic!

Cybersecurity certifications act as a testament to your knowledge and skills in the field. They not only validate your expertise but also showcase your dedication toward continuous learning, a trait highly valued in the ever-evolving world of cybersecurity. Certifications can help bridge knowledge gaps, provide you with a competitive edge, and are sometimes even a prerequisite for certain roles.

Let's dive into a variety of certifications in cybersecurity, ranging from beginner to advanced levels.

Selecting a certification depends on your current knowledge level, your career aspirations, and the specific area of cybersecurity you're interested in. Remember, a certification alone won't make

CERTIFICATION	DESCRIPTION	WHO IS IT FOR	DIFFICULTY LEVEL
CompTIA Security+	A global certification that validates baseline skills necessary to perform core security functions and pursue an IT security career.	Entry-level professionals	Beginner
Certified Ethical Hacker (CEH)	This certifies your knowledge in finding weaknesses and vulnerabilities in target systems, using the same tools as a malicious hacker, but lawfully and legitimately.	Professionals who want to be ethical hackers	Intermediate
Certified Information Systems Security Professional (CISSP)	Recognized globally as a standard of achievement, it validates managerial competence in overseeing and governing an enterprise's information security program.	Professionals aiming for a senior role in IT security	Advanced
Certified Information Systems Auditor (CISA)	This certification validates your knowledge for information systems auditing, control, and security.	Professionals who conduct information systems audits	Intermediate

(*Continued*)

CERTIFICATION	DESCRIPTION	WHO IS IT FOR	DIFFICULTY LEVEL
Certified Information Security Manager (CISM)	This certification is focused on the management and governance of IT security and is geared toward professionals involved in the development and management of an enterprise information security program.	IT security management professionals	Advanced
GIAC Certified Incident Handler (GCIH)	The GCIH certifies a professional's ability to manage incidents, to understand common attack techniques and responses, and the legal issues around incidents.	Professionals in incident handling and defending against cyber threats	Advanced
Offensive Security Certified Professional) (OSCP)	This is a rigorous and respected certification that validates your ability to perform a penetration test using a methodological approach. It's mostly hands-on and, hence, highly practical.	Security professionals involved in penetration testing and vulnerability assessment	Highly Advanced

you a cybersecurity expert overnight. However, the learning journey you undertake while pursuing these certifications can significantly enhance your understanding of the field, making you a better cybersecurity professional.

6.2 Key Sections for a Cybersecurity Resume

- **Summary:** This brief overview highlights your key qualifications and career aspirations. It should give recruiters a quick understanding of what you bring to the table.
- **Education:** List your degrees, the institutions you attended, and the years of your attendance.
- **Experience:** Highlight your previous roles, responsibilities, and achievements. Be sure to use active voice and include metrics where applicable.
- **Tech Stack:** Outline the technical tools, languages, and systems you're proficient in.

- **Certifications:** Include any relevant certifications you've earned, such as Certified Information Systems Security Professional (CISSP) or Certified Information Security Manager (CISM).
- **Open-Source Projects:** If you have contributed to any open-source projects or have undertaken any individual projects relevant to the role, include them here. This can demonstrate your practical skills and initiative.

Remember, your resume is your advocate during the recruitment process. Make sure it is an accurate and compelling reflection of your abilities, showcasing you as an ideal candidate for the role.

6.3 Do's and Don'ts for Resume Writing

Do's:

✓ Use relevant keywords and phrases from the job description throughout your resume.
✓ Include numbers, metrics, and quantifiable achievements like "Increased website traffic by 20%."
✓ Structure your tech skills section with columns for proficiency level, years of experience, certifications.
✓ Focus on technical skills, tools, and programming languages used on the job.
✓ Tailor your resume to each application – emphasize skills needed for that role.
✓ Use a clean, simple format with scannable headers and clear sections.

Don'ts:

✖ Use resume templates with complex formatting, graphics, or colors.
✖ Include soft skills like "team player" without tangible examples.
✖ Use dense blocks of text – apply bullet points to help readability.
✖ List basic skills like MS Office programs unless highly specialized.
✖ Leave gaps in employment – explain sabbaticals, family leave, etc.

- ✖ Exaggerate or lie about your skills and experience.
- ✖ Use personal pronouns like "I" or include photos.
- ✖ List irrelevant hobbies, interests, or social media handles.
- ✖ Forget to customize your resume for each application.

6.4 Sample Resume

John Doe

123 Main Street, City, State 12345 | (123) 456-7890 | john.doe@email.com

6.5 Summary

Motivated cybersecurity analyst with a bachelor's degree in information technology and a Certified Information Systems Security Professional (CISSP). Skilled in identifying and mitigating cyber vulnerabilities, maintaining data integrity, and performing risk assessments. Passionate about evolving security trends and dedicated to contributing toward a safer digital world.

6.5.1 Education

Bachelor of Science in Information Technology, XYZ University, City, State | August 2020 – May 2023

6.5.2 Certifications

Certified Information Systems Security Professional (CISSP) | June 2023

Skills

- Vulnerability assessment and mitigation
- Intrusion detection and prevention systems (IDS/IPS)
- Network security
- Knowledge of TCP/IP protocols
- Familiarity with firewalls, VPN, SIEM
- Programming: Python, Java

6.5.3 Experience

Cybersecurity Intern | ABC Tech, City, State | June 2023 – Present

- Assisted in managing the implementation of security solutions.
- Participated in the development of incident response plans and conducted a successful simulated security event.
- Identified and addressed malware and malicious activities in the company network.
- Assisted in the analysis of security systems and upgrade recommendations.
- Prepared security reports to track incidents and events.

6.5.4 Open-Source Projects

SecureHome: Developed a home security system using Raspberry Pi that alerts homeowners of any breaches through SMS | GitHub Link: GitHub.com/SecureHome

6.5.5 Languages

Python, Java, SQL

Remember, the above is just a sample. Your resume may look different based on your own unique experiences, skills, and qualifications. Be sure to personalize your resume to reflect your own background and the specific job requirements.

6.6 Conclusion

Pursuing relevant certifications and properly representing your skills and experiences on your resume are important steps for building a successful cybersecurity career. Certifications validate and strengthen your expertise, while an effective resume markets you as a qualified candidate to hiring managers. Tailor your certifications and resume to match your current skill level, interests, and the specific roles you are targeting. Cybersecurity is a continuously evolving field that values up-to-date knowledge, practical abilities, and a dedication to lifelong learning. With the right certifications and resume, you can demonstrate your commitment to the field and enhance your competitiveness as a cybersecurity professional. The key is to showcase your qualifications accurately while highlighting your passion to grow within this dynamic and critical industry.

Knowledge Check

Question 1: What does using active voice and keywords helps make your resume?
Answer: More compelling.

Question 2: What are two important sections to include on a cybersecurity resume?
Answer: Skills and certifications.

Question 3: True or False: Your resume should be generic and not tailored for each role.
Answer: False.

7

Understanding the Recruitment Process

Breaking into the field of cybersecurity involves navigating through the recruitment process. While every company's process can be slightly different, the stages often include an initial screening, technical and behavioral interviews, and, finally, an offer.

One crucial point to remember during the recruitment process is that companies typically focus on three core aspects while hiring:

- **Solving Immediate Problems:** Companies often have immediate cybersecurity needs, and they're looking for professionals who can hit the ground running. Whether it's fortifying their defenses, responding to an ongoing threat, or complying with new regulations, your ability to step in and address these needs can make you stand out as a candidate. Demonstrating your practical knowledge of cybersecurity concepts, tools, and best practices during the interview can reassure employers of your capability to solve their immediate problems.
- **Adding Value:** Beyond the immediate problems, companies are looking for individuals who can add value to their organization in the long term. This could involve developing new security protocols, training other staff members, or contributing to the overall cybersecurity strategy. Highlighting your potential contributions can make you an attractive candidate.
- **Return on Investment:** Hiring is an investment, and companies are looking for a high return. If a company is going to invest $100,000 in your salary, they want to see that you're bringing far more value to the organization. This is where your unique blend of skills, experiences, and potential growth come into play. Illustrating how your expertise will drive efficiency, reduce risks, or save costs can position you as a high-value candidate.

DOI: 10.1201/9781003501589-8

Keep these three points in mind as you prepare for interviews. Showcase not only your knowledge of cybersecurity but also your understanding of how you can apply that knowledge to benefit the business. By aligning your skills and experiences with the company's needs, you'll put yourself in a strong position to secure your first job in cybersecurity.

Note

Succeeding in cybersecurity job interviews involves more than just demonstrating technical prowess. While showcasing your hard skills is crucial, you must also exhibit professional polish and interpersonal abilities. Here are some additional tips for navigating the recruitment process smoothly:

- Thoroughly research the company and position ahead of time to articulate how your background aligns with their needs. This preparation conveys genuine interest.
- Ask thoughtful questions that reveal strategic perspective, not just technical curiosities. For example, inquire about balancing security and usability in products.
- Have stories ready that highlight soft skills like communication, teamwork, and adaptability. Technical expertise is expected, but personable collaboration is invaluable.
- Keep answers positive, even when discussing previous challenges. Focus on acquired skills rather than negativity.
- Follow up promptly with thank you notes after each interview to reiterate enthusiasm.
- Be ready to explain any gaps or shifts in your background as growth opportunities.
- Treat each interviewer, regardless of level, with equal respect. You never know who has influence.
- Before leaving, ask about next steps to display preparation for advancing.

With the right balance of technical knowledge and interpersonal skills, you can make an outstanding impression throughout the cybersecurity hiring process.

Now, let's look at some specific cybersecurity job titles with sample questions and answers.

JOB TITLE	INTERVIEW QUESTION	SAMPLE ANSWER
Information Security Analyst	1. What is a SIEM system and how does it work?	SIEM stands for Security Information and Event Management. It is a technology that aggregates and analyzes activity from various resources across an entire IT infrastructure. SIEM collects security data from network devices, servers, domain controllers, and more. It allows for threat detection, forensic analysis, and incident response by identifying patterns and anomalous activity.
	2. Can you describe a recent vulnerability you discovered and how you addressed it?	In my previous role, I detected an SQL Injection vulnerability on our company's website. I alerted our web development team immediately and advised them on how to patch the vulnerability. I also recommended conducting a full review of our site's code to ensure no similar vulnerabilities existed.
	3. How do you stay updated on the latest security threats and data breaches?	I subscribe to various cybersecurity newsletters and blogs like KrebsOnSecurity, Schneier on Security, and the SANS Institute. I also follow cybersecurity experts on social media and participate in relevant forums and webinars.
Network Security Engineer	1. How do you secure a network?	Securing a network involves multiple layers of protection. First, I ensure all systems, software, and firmware are updated regularly. Next, I enforce the use of strong passwords and two-factor authentication. I also implement firewalls, intrusion detection systems, and antivirus software. Regular audits and backups are also critical.
	2. Explain how a three-way handshake works.	A three-way handshake is a process used in TCP/IP networks to create a connection between a client and server. It involves three steps: (1) The client sends a SYN packet to the server. (2) The server acknowledges the client's SYN packet by sending a SYN-ACK packet back. (3) The client sends an ACK packet back to the server, and the connection is established.
	3. How would you handle a DDoS attack?	My first step would be to identify the nature of the attack, such as volumetric, protocol, or application layer. Then, I'd apply the appropriate mitigation strategies, like rate limiting, IP blocking, or request filtering. I would also coordinate with our ISP or DDoS protection service for additional mitigation. Post-attack, I'd conduct a thorough analysis to learn from the event.

(*Continued*)

JOB TITLE	INTERVIEW QUESTION	SAMPLE ANSWER
Penetration Tester	1. What are the key stages of a penetration test?	Penetration testing generally involves five stages: (1) Planning and reconnaissance where you define the scope and gather information; (2) Scanning, where you understand how the target application or system responds to various intrusion attempts; (3) Gaining access, which involves web application attacks; (4) Maintaining access, to see if the vulnerability can be used to achieve a persistent presence in the exploited system; (5) Analysis and reporting, to compile test results and suggest countermeasures.
	2. Can you describe a time when you identified and exploited a security vulnerability?	In my previous role, I discovered a Cross-Site Scripting (XSS) vulnerability during a penetration test on our client's e-commerce website. I was able to inject a malicious script into the website's checkout page. I reported the issue, and we worked closely with the client to resolve the issue promptly.
	3. How familiar are you with the OWASP Top 10?	I'm very familiar with the OWASP Top 10. They are a powerful awareness tool for web application security that represents a broad consensus about the most critical security risks. I regularly use the OWASP Top 10 as a guide when conducting web application penetration testing and vulnerability assessments.
Incident Responder	1. How would you handle a suspected malware infection?	I would first isolate the affected system to prevent further spread of the malware. Then, I would conduct an investigation to identify the type of malware, how it got in, and the extent of the damage. Once I've collected this information, I would remove the malware, restore the system, and take steps to prevent similar infections in the future.
	2. Can you describe the steps in an incident response plan?	An incident response plan typically involves five stages: (1) Preparation: having the right tools, team, and plan in place; (2) Identification: determining whether an incident has occurred; (3) Containment: limiting the scope and magnitude of the incident; (4) Eradication: removing the cause of the incident and any damage; (5) Recovery: restoring systems to normal operations and confirming the systems are no longer compromised.
	3. What would you do after a security breach?	I would immediately start an investigation to understand what happened, document the incident and the steps taken, ensure the vulnerability is patched, and report to the relevant stakeholders, possibly including customers if their data was compromised.

(*Continued*)

JOB TITLE	INTERVIEW QUESTION	SAMPLE ANSWER
Security Architect	1. What principles do you consider when designing a secure network architecture?	A secure network should be designed with defense in depth, least privilege, network segmentation, redundancy, and secure configurations.
	2. How would you balance business needs with security?	It's crucial to understand that security exists to enable the business, not restrict it. I would work closely with business stakeholders to understand their needs and then design a secure solution that meets those needs while minimizing risk.
	3. Can you describe a time when you had to adapt your design due to constraints?	During a project at my previous job, budget constraints prevented us from implementing a proposed security solution. Instead of insisting on the original plan, I reassessed the situation and found a less expensive, but still effective solution that fit our needs.
Malware Analyst	1. How do you analyze malware?	I typically start by creating a secure and isolated lab environment to prevent accidental infections. Then I use static analysis techniques to examine the malware without running it and dynamic analysis techniques to observe how the malware behaves when run.
	2. Can you discuss a recent malware threat you've dealt with?	I recently analyzed a ransomware variant that utilized double encryption, making it significantly more challenging to decrypt and recover the files. It was a stimulating and rewarding experience that allowed me to enhance my decryption skills.
	3. How do you stay up-to-date with the latest malware threats?	I regularly read cybersecurity news, research papers, and blogs. I'm also a member of several online communities where we share information about new threats.
Cybersecurity Consultant	1. How would you evaluate a client's security posture?	I would conduct a thorough assessment, which could include a review of their security policies, vulnerability assessments, penetration testing, and evaluation of their incident response capabilities.
	2. Can you give an example of a security improvement you've recommended to a client?	I once recommended that a client adopt two-factor authentication across all their systems. They had experienced a series of phishing attacks, and this extra layer of security significantly reduced the risk of account compromise.
	3. How do you handle pushback from a client?	Clear communication and education are essential. I strive to explain my recommendations in a way that demonstrates the value and necessity of the proposed security measures.

(*Continued*)

JOB TITLE	INTERVIEW QUESTION	SAMPLE ANSWER
Ethical Hacker	1. What's the difference between white box, black box, and gray box testing?	White box testing is when the tester has full knowledge of the system being tested. Black box testing is when the tester has no knowledge of the system. Gray box is a mix of both, where the tester has limited knowledge of the system.
	2. Can you describe a time when you found a critical vulnerability during a penetration test?	During a recent engagement, I discovered a server that was misconfigured to allow unrestricted file uploads. This could have allowed an attacker to upload a malicious script and gain control of the server.
	3. What tools do you commonly use in your work?	I use a variety of tools depending on the task at hand. Some of my go-to's include Metasploit for developing and executing exploit code, Wireshark for packet analysis, and Burp Suite for web application testing.
Cybersecurity Analyst	1. What is the role of a cybersecurity analyst?	A cybersecurity analyst is responsible for protecting both company networks and data. This includes threat hunting, responding to incidents, analyzing security systems, and recommending improvements.
	2. How would you handle a phishing attack?	I would first confirm the attack and identify the affected users. Then I would remove the phishing emails from the users' inboxes, change their passwords, and scan their systems for malware. I would also report the phishing attack to the appropriate authorities and educate the users on how to recognize phishing emails.
	3. Can you describe a time when you successfully mitigated a threat?	At my last job, I detected unusual network traffic that turned out to be an attempted data breach. I immediately isolated the affected systems and helped remove the malicious software. I then worked with the team to strengthen our security measures and prevent similar attacks.
Cloud Security Engineer	1. How do you secure data in the cloud?	Data in the cloud should be protected at rest, in transit, and during processing. Encryption is critical for protecting data at all stages. I also consider the security of the cloud service provider and follow best practices for access control and configuration.
	2. What are some challenges of cloud security?	Some of the key challenges include managing access controls, dealing with the shared responsibility model, maintaining privacy, and dealing with regulatory compliance.

(*Continued*)

JOB TITLE	INTERVIEW QUESTION	SAMPLE ANSWER
	3. How do you monitor security in a cloud environment?	I use a combination of log analysis, intrusion detection systems, and security management tools provided by the cloud service provider. Regular audits are also important for ensuring security controls are working as intended.
Chief Information Security Officer (CISO)	1. How do you align security strategy with business objectives?	Security should enable the business, not hinder it. I work closely with other executives to understand the business goals and develop a security strategy that supports these goals while managing risk.
	2. How do you foster a security-conscious culture in an organization?	This requires a combination of training, awareness campaigns, and making security part of everyday processes. It's also important to get buy-in from the top levels of the organization.
	3. How do you handle budget constraints?	I prioritize spending based on risk. Not all risks can be eliminated, but they can be managed. I focus on the biggest risks and the most effective controls, and I always look for cost-effective solutions.
Forensic Computer Analyst	1. Describe a challenging case you worked on.	I worked on a case involving an insider threat where an employee was suspected of stealing sensitive data. I had to carefully analyze the employee's actions on the network while ensuring the integrity of the evidence.
	2. What tools do you use for digital forensics?	I use tools like EnCase and FTK for disk and data analysis, Wireshark for network forensics, and Volatility for memory forensics.
	3. How do you ensure the integrity of your digital evidence?	I follow strict chain of custody procedures, make bit-by-bit copies of digital media for analysis, and use hashing algorithms to verify that the data has not been altered.
Entry-Level Security Analyst	1. How would you respond if you detected an unusual activity on the network?	I would first document the anomaly, then investigate further to confirm whether it's a false positive or a genuine security event. Depending on the severity, I might need to escalate the issue according to the organization's incident response plan.
	2. How familiar are you with cybersecurity frameworks?	I have a strong understanding of several cybersecurity frameworks such as NIST, ISO 27001, and CIS. These frameworks guide organizations in managing and reducing their risk related to cybersecurity.
	3. Can you describe a time you helped solve a technical problem?	In a previous role, I assisted in troubleshooting a recurring network issue that was affecting productivity. By carefully analyzing the network logs, I was able to identify the source of the problem and suggest a solution.

(*Continued*)

JOB TITLE	INTERVIEW QUESTION	SAMPLE ANSWER
Junior Penetration Tester	1. What methodology do you follow when conducting a penetration test?	I follow the standard steps of penetration testing: planning and reconnaissance, scanning, gaining access, maintaining access, and analysis/reporting. Each step is crucial to a successful test.
	2. Can you explain what a cross-site scripting attack is?	A cross-site scripting (XSS) attack occurs when an attacker injects malicious script into web pages viewed by other users. These scripts can then be used to steal sensitive data or perform other malicious actions.
	3. What are some tools you've used for penetration testing?	I've used tools like Metasploit for developing and executing exploit code, Wireshark for packet analysis, and OWASP ZAP for web application testing.
IT Auditor	1. How do you approach an IT audit?	The first step is to understand the audit's scope and objectives. Then, I would review the organization's IT policies, procedures, and controls to evaluate their effectiveness. I would also conduct interviews with key personnel and perform testing as needed.
	2. Can you describe an example of a control you might evaluate during an IT audit?	One control I might evaluate is the process for managing user access to systems. This would involve reviewing how access is granted, how rights are modified or revoked, and how regularly access rights are reviewed and confirmed.
	3. What standards or frameworks do you follow when conducting an IT audit?	The Control Objectives for Information and Related Technologies (COBIT) framework is one standard I follow for IT governance and management. I also reference the ISO 27001 standard for information security management.
Risk Analyst	1. How do you assess risk?	I use a risk assessment framework, which involves identifying potential threats, determining the vulnerabilities that could be exploited by those threats, estimating the potential impact, and prioritizing the risks based on their likelihood and impact.
	2. Can you explain the concept of risk appetite?	Risk appetite is the level of risk that an organization is willing to accept in pursuit of its objectives. It varies from one organization to another and should align with the organization's strategy and culture.
	3. How do you communicate risk to stakeholders?	It's important to communicate risk in a way that is easy for nontechnical stakeholders to understand. I try to use clear, concise language and avoid technical jargon. Visual aids like risk heat maps can also be helpful.

(*Continued*)

JOB TITLE	INTERVIEW QUESTION	SAMPLE ANSWER
Compliance Analyst	1. How do you ensure an organization stays compliant with laws and regulations?	I keep up-to-date with relevant laws and regulations, conduct regular audits to check for compliance, and work with other departments to implement policies and procedures that promote compliance.
	2. Can you explain the concept of a compliance framework?	A compliance framework is a structured set of guidelines that details an organization's processes for adhering to legal, security, and industry-specific rules and regulations.
	3. How would you handle a potential violation of compliance standards?	I would document the potential violation, conduct an investigation to understand what happened, and work with relevant teams to correct the issue. I would also review our policies and procedures to see if changes are needed to prevent similar violations in the future.
Junior GRC Analyst	1. Can you explain the three elements of GRC and why they are important?	Governance, Risk, and Compliance (GRC) are three interrelated facets of corporate strategy. Governance refers to the overall management approach of the organization. Risk involves identifying, assessing, and managing potential disruptions to the organization's objectives. Compliance ensures the organization follows laws, regulations, standards, and ethical practices. Together, they provide a comprehensive approach to organizational leadership and decision-making.
	2. What are some common GRC tools you have experience with?	I have worked with tools like RSA Archer and MetricStream, which provide integrated solutions for managing all aspects of GRC.
	3. How would you assist in a GRC implementation project?	I would start by understanding the organization's objectives and current processes. Then, I would assist in designing a GRC strategy that aligns with those objectives, select a suitable GRC tool, and work with different teams to implement the strategy.

7.1 Conclusion

Navigating the recruitment process effectively is vital for launching a successful career in cybersecurity. Technical expertise alone is not enough – you must also demonstrate aptitude in areas like communication, strategic thinking, and cultural alignment. Thorough preparation including researching the company, practicing responses, and asking insightful questions can help you stand out. Exhibiting enthusiasm, respect, and professionalism in all interactions builds a positive impression. With both hard and soft skills finely tuned, you can show how your unique background solves immediate problems and adds long-term value, resulting in a great return on the company's hiring investment. Keep focused on conveying your technical capabilities while also highlighting your collaboration abilities and growth potential. With the right balance, you can excel in cybersecurity interviews and recruitment to land your ideal role.

Quizzes

Quiz 1: Question: What are the three traits companies look for in cybersecurity candidates?
Answer: Ability to solve immediate problems, add long-term value, provide high ROI.

Quiz 2: Question: Soft skills like communication and collaboration are __ in interviews.
Answer: Just as important as hard skills.

Quiz 3: Question: True or False: Asking insightful questions in interviews demonstrates interest.
Answer: True.

8

When You Get the Offer

Next Steps

So, after a series of interviews and rigorous selection procedures, you've finally got an offer for a cybersecurity job. Congratulations! But, hold your horses; the journey isn't over just yet. The period that follows the job offer is equally crucial, involving critical decision-making, negotiations, and setting the right expectations. Let's unpack this process.

1 **Negotiation: An Expected Part of the Process**
It's important to understand that negotiations are an expected part of the recruitment process. No matter how tempting it might be to accept the first offer that comes your way, it's usually a good idea to pause and evaluate. Is this the best offer for you? Could you potentially negotiate for better terms?

2 **Salary and Time Off: Key Areas for Negotiation**
Salary: Salary is often the first point of negotiation. Have a clear understanding of what your skills and experience are worth in the market. There are numerous resources online to help you gauge an acceptable salary range for your role. If the proposed salary seems low, don't be afraid to ask for more. Remember, it's not just about the number – it's about being compensated fairly for the value you bring to the team.

Time Off: Time off is another area of negotiation. Consider your personal needs and whether the offered vacation time meets those needs. If not, negotiate for more. This includes discussing public holidays, sick leaves, and personal time off.

3 **Bonuses, Benefits, and Work Schedule**
Finally, sign-on bonuses, special benefits, and remote work schedule are other factors that can also be negotiated. **Sign-on bonuses** can help cushion the transition between jobs. **Special benefits** can range from health care plans, retirement plans,

DOI: 10.1201/9781003501589-9

educational assistance, to gym memberships and beyond. As for the **work schedule**, with the pandemic having normalized remote work, you might want to negotiate for a flexible schedule that can accommodate remote working days.

Remember, The key to a successful negotiation lies in respectful communication, backed by reasonable justifications for your requests. Always be prepared to explain why you think you deserve what you're asking for.

Once you've negotiated your terms and are satisfied with the offer, it's time to **accept and prepare for the exciting journey ahead**.

8.1 Carefully Reviewing the Offer Details

You've negotiated an offer, but don't sign just yet! Before accepting, be sure to carefully evaluate all aspects of the offer letter and employment contract. Scrutinize the fine print details and any areas that seem unclear or concerning. This step is essential to avoid any misunderstandings about the actual terms of employment.

8.2 Conclusion

This concludes our journey through this book. You are now armed with a comprehensive understanding of the cybersecurity landscape, the role of IT in businesses, and the skills needed to land your first cybersecurity job. My aim was to provide practical tips and real-world advice and to make this guide your stepping stone toward a rewarding career in cybersecurity.

Remember, the world of cybersecurity is constantly evolving, and so should you. Keep learning, stay curious, and never become complacent in your skills. Attend training, read industry resources, connect with peers, and get certified in new technologies.

View obstacles as opportunities to grow and refine your expertise. With perseverance and dedication, you can turn setbacks into springboards.

Finally, don't just strive to conquer the cybersecurity world alone. Collaborate with and mentor others entering the field. Share your knowledge and lift up those around you. The future of cybersecurity depends on collective vigilance, teamwork, and a commitment to excellence.

You now have the foundation needed to make a real impact. I wish you the very best as you embark on an exciting and critically important career protecting the digital world we all rely on. Onward!

Quizzes

Quiz 1: Question: Salary, time off, and bonuses are all standard areas of _____.
Answer: Negotiation.

Quiz 2: Question: Before accepting an offer, you should _____.
Answer: Thoroughly review all details.

Quiz 3: Question: True or False: You should accept the first offer you receive without negotiation.
Answer: False.

Post-Assessment Quiz

1. What are the three aspects of the CIA triad?
   (a) Cryptography, Identity, Access
   (b) Confidentiality, Integrity, Availability
   (c) Compliance, Integration, Awareness
   (d) Confidentiality, Internet, Authentication
2. What are the stages of an incident response plan?
   (a) Identification, Containment, Mitigation
   (b) Preparation, Detection, Recovery
   (c) Prevention, Response, Diagnosis
   (d) Planning, Implementation, Evaluation
3. What does DLP stand for?
   (a) Data Loss Prevention
   (b) Distributed Link Protocol
   (c) Data Leakage Protection
   (d) Dual-Layer Protocol
4. What is ransomware?
   (a) Malware designed to gain network access
   (b) Software that holds systems hostage for ransom
   (c) Viruses that exploit operating system vulnerabilities
   (d) Worms that breach firewalls and propagate

5. What is a proxy server?
   (a) A network gateway that filters web traffic
   (b) A protocol that enables remote desktop access
   (c) An attack vector used to intercept network traffic
   (d) A virtual server only accessible on the deep web
6. What is a penetration test?
   (a) Testing defenses by simulating cyberattacks
   (b) Developing malicious software and exploitation tools
   (c) Attempting to steal data and compromise security
   (d) Publishing confidential information publicly
7. What is reconnaissance in cybersecurity?
   (a) Retrieving stolen data from adversaries
   (b) Identifying target vulnerabilities and weaknesses
   (c) Disabling firewalls and other security controls
   (d) Gaining remote access to systems using malware
8. What is a security operations center (SOC)?
   (a) A team that responds to security incidents
   (b) A network operations center focused on uptime and performance
   (c) A server room where log data is aggregated and stored
   (d) A virtual environment used to safely analyze malware
9. What is threat intelligence?
   (a) Using social engineering to obtain sensitive information
   (b) Gathering and analyzing data to understand adversaries
   (c) Attempting to compromise or disable target systems
   (d) Stealing and publishing confidential data publicly
10. What is multifactor authentication?
    (a) A password consisting of multiple words
    (b) Authentication using multiple approaches like passwords and biometrics
    (c) Granting system access permissions to privileged users
    (d) Encrypting data across multiple algorithms

Answers to Quizzes

Alright, cyber warriors! You've made it through the battleground of quizzes in "Zero to Hero: Your Guide to a Career in Cybersecurity." Now, it's time to see how well you've really done. This Answer Reference Guide is your secret weapon—a cheat code if you will—to double-check your answers and ensure you're on the right track.

Think of it as your trusty sidekick, ready to swoop in and confirm that you've nailed those tricky questions. Whether you're here to pat yourself on the back or catch those pesky little mistakes, this guide's got your back. So, dive in, compare your answers, and keep leveling up your cybersecurity game!

Chapter 1: Understanding Businesses: What and Why

1 What is the main reason businesses exist?
Answer: To make a profit.
2 How do businesses generate revenue?
Answer: By selling goods or services.
3 What is the difference between revenue and profit?
Answer: Revenue is total income before expenses. Profit is revenue minus expenses.

Chapter 2: The Role of IT in Business

1. How has IT impacted modern businesses?
   Answer: Made them more efficient, interconnected, and globally competitive.
2. Name two ways IT can help businesses increase revenue.
   Answer: E-commerce platforms and cloud-based subscription services.
3. What are two cybersecurity risks introduced by reliance on IT?
   Answer: Data breaches and disruption of operations from cyberattacks.

Chapter 3: The Pillars of IT and Cybersecurity

1. As dependence on IT grew, what emerged in parallel?
   Answer: Cyber threats.
2. What are the three pillars of IT evolution?
   Answer: Hardware, Networking, Databases.
3. Name one implication of cyber threats for businesses.
   Answer: Financial loss, reputational damage, disruption of operations.

Chapter 4: Simplifying Cybersecurity: Key Areas and Concepts

1. What are the three aspects of the CIA triad?
   Answer: Confidentiality, Integrity, Availability.
2. What do the four steps of IAAA stand for?
   Answer: Identification, Authentication, Authorization, Accountability.
3. Name one popular cybersecurity framework.
   Answer: NIST, ISO 27001, COBIT, CIS.

Chapter 5: Preparing for the Job Hunt: What You Need to Know

1. What are the three main stages of a cybersecurity job interview process?
   Answer: Screening, Technical Interview, Final Interview.
2. What two types of interview questions should you expect and prepare for?
   Answer: Scenario-based and Technical.

3 True or False: You should only answer interview questions, not ask your own.
Answer: False.

Chapter 6: Mastering Your Resume

1 What does using active voice and keywords help make your resume?
Answer: More compelling.
2 What are two important sections to include on a cybersecurity resume?
Answer: Skills and certifications.
3 True or False: Your resume should be generic and not tailored for each role.
Answer: False.

Chapter 7: Understanding the Recruitment Process

1 What are the three traits companies look for in cybersecurity candidates?
Answer: Ability to solve immediate problems, add long-term value, provide high ROI.
2 Soft skills like communication and collaboration are __ in interviews.
Answer: Just as important as hard skills.
3 True or False: Asking insightful questions in interviews demonstrates interest.
Answer: True.

Chapter 8: When You Get the Offer: Next Steps

1 Salary, time off, and bonuses are all standard areas of ___.
Answer: Negotiation.
2 Before accepting an offer, you should ___.
Answer: Thoroughly review all details.
3 True or False: You should accept the first offer you receive without negotiation.
Answer: False.

Post-Assessment Quiz

1. What are the three aspects of the CIA triad?
   Answer: Confidentiality, Integrity, Availability.
2. What are the stages of an incident response plan?
   Answer: Preparation, Detection, Recovery.
3. What does DLP stand for?
   Answer: Data Loss Prevention.
4. What is ransomware?
   Answer: Malware that holds systems hostage for ransom.
5. What is a proxy server?
   Answer: A network gateway that filters web traffic.
6. What is a penetration test?
   Answer: Testing defenses by simulating cyberattacks.
7. What is reconnaissance in cybersecurity?
   Answer: Identifying target vulnerabilities and weaknesses.
8. What is a security operations center (SOC)?
   Answer: A team that responds to security incidents.
9. What is threat intelligence?
   Answer: Gathering and analyzing data to understand adversaries.
10. What is multifactor authentication?
    Answer: Authentication using multiple approaches like passwords and biometrics.

Index

Pages in **bold** refer to tables

T

V

W

Printed in the United States
by Baker & Taylor Publisher Services